BSCS Science Tracks

Connecting Science and Literacy

Investigating
Weather

BSCS

KENDALL/HUNT PUBLISHING COMPANY
4050 Westmark Drive   Dubuque, Iowa 52002

Lance Johnson, Grade 5, Memorial Intermediate, Fitchburg, MA

Kyle Kimmal, Grade 3, Green Valley Elementary School, Denver, CO

Maureen Lampa, Grade 3, Reingold Elementary School, Fitchburg, MA

Jennifer Levine, Grade 3, Bay Park Elementary School, San Diego, CA

Pam Magill, Grade 3, Steck Elementary School, Denver, CO

Elizabeth Maki, Grade 3, South Street Elementary School, Fitchburg, MA

Jeannette Martinez, Grade 4, South Street Elementary School, Fitchburg, MA

Lauren McKittrick, Grade 3, Green Valley Elementary School, Denver, CO

Samantha Messier, Denver Public Schools, Denver, CO

Serri Mills, Grade 5, Park Hill Elementary School, Denver, CO

Andrea Oulette, Grade 4, Crocker Elementary School, Fitchburg, MA

Tammi Parisi, Grade 4, Reingold Elementary School, Fitchburg, MA

Carole Pierce, Grade 4, Doull Elementary School, Denver, CO

Marisa Ramirez, San Diego City Schools, San Diego, CA

Soledad Ramirez-Heiler, Grade 4, Wexford Elementary School, Lansing, MI

Christy Ranbarger, Grade 3, Bay Park Elementary School, San Diego, CA

Joan Romano, Grade 4, Reingold Elementary School, Fitchburg, MA

Marci Roy, Grade 3, Reingold Elementary School, Fitchburg, MA

Sabra Scheel, Grade 4, Maple Grove Elementary School, Lansing, MI

Jason Shiroff, Grade 4, Southmoor Elementary School, Denver, CO

Jim Short, Denver Public Schools, Denver, CO

R. Timothy Smith, Lansing School District, Lansing, MI

Catie Solan, Grade 3, Reingold Elementary School, Fitchburg, MA

Hilary Spark, Grade 5, Memorial Intermediate, Fitchburg, MA

Leslie Stahl, Grade 3, Southmoor Elementary School, Denver, CO

Esther Supinski, Grade 3, Crocker Elementary School, Fitchburg, MA

Glenda Swazo, Grade 4, Green Valley Elementary School, Denver, CO

Carrie Symons, Grade 4, Harrington Elementary School, Denver, CO

Randy Jo Thomas, Grade 3, Southmoor Elementary School, Denver, CO

Maria C. Torres, Grade 3, Reingold Elementary School, Fitchburg, MA

Kimberly Toupin, Grade 3, Crocker Elementary School, Fitchburg, MA

Bonnie Walters, Denver Public Schools, Denver, CO

Deborah Welch, Grade 3, Reingold Elementary School, Fitchburg, MA

Connie Wilson, Grade 4, Forest View Elementary School, Lansing, MI

Cindy Wironen, Grade 3, Reingold Elementary School, Fitchburg, MA

BSCS Staff, First Edition

BSCS Development Team

Nancy M. Landes, Project Director and Author, 1996–1998

Gail C. Foster, Author

Colleen K. Steurer, Author

Vonna G. Pinney, Executive Assistant

Linda K. Ward, Senior Executive Assistant

Rodger W. Bybee, Principal Investigator, 1994–1995

Harold Pratt, Project Director, 1994–1996

Janet Chatlain Girard, Art Coordinator, 1994–1996

BSCS Administrative Staff

Timothy H. Goldsmith, Chair, Board of Directors

Joseph D. McInerney, Director

Michael J. Dougherty, Assistant Director

Lynda B. Micikas, Assistant Director

Larry Satkowiak, Chief Financial Officer

Contributors and Consultants

Randall K. Backe, BSCS, Colorado Springs, Colorado, Contributing Author

Judy L. Capra, Wheat Ridge, Colorado, Freelance Writer

Michael J. Dougherty, BSCS, Colorado Springs, Colorado

B. Ellen Friedman, San Diego, California
Cathy Griswold, Lyons, Oregon, Contributing Author
Jay Hackett, Greeley, Colorado
Debra A. Hannigan, Colorado Springs, Colorado, Contributing Author
David A. Hanych, BSCS, Colorado Springs, Colorado
Karen Hollweg, Washington, DC
Winston King, Bridgetown, Barbados
Paul Kuerbis, Colorado Springs, Colorado
Donald E. Maxwell, BSCS, Colorado Springs, Colorado
Brenda S. McCreight, Colorado Springs, Colorado, Freelance Writer
Mary McMillan, Boulder, Colorado
Marge Melle, Littleton, Colorado, Freelance Writer
Lynda B. Micikas, BSCS, Colorado Springs, Colorado
Jean P. Milani, BSCS, Colorado Springs, Colorado
Renee Mitchell, Lakewood, Colorado, Freelance Writer
Janet Carlson Powell, Boulder, Colorado, Contributing Author
Carol D. Prekker, Broomfield, Colorado, Freelance Writer
Patricia J. Smith, Tucson, Arizona, Freelance Writer
Terry G. Spencer, Monterey, California, Contributing Author
Patti M. Thorn, Austin, Texas, Contributing Author
Bonnie Turnbull, Monument, Colorado, Freelance Writer
Terri B. Weber, Colorado Springs, Colorado
Carol A. Nelson Woller, Boulder, Colorado, Freelance Writer

Field-Test Teachers and Coordinators

Joanne Allen, Westport Elementary School, Westport, Maine
Helene Auger, Westport School District, Westport, Maine
Sheila Dallas, Bethany School, Cincinnati, Ohio
Pat Dobosenski, Pembroke Elementary School, Troy, Michigan
Nina Finkel, Whitter Elementary School, Chicago, Illinois
Mary Elizabeth France, Westport Elementary School, Westport, Maine
Carolyn Gardner, Calhan Elementary School, Calhan, Colorado

Shelly Gordon, Bingham Farms Elementary School, Birmingham, Michigan
Darlene Grunert, Birmingham Public Schools, Birmingham, Michigan
Terry Heinecke, Edgerton Elementary School, Kalispell, Montana
Katherine Hickey, Irving Primary School, Highland Park, New Jersey
Jan Himmelspach, Grayson Elementary School, Waterford, Michigan
Elizabeth Lankes, Bethany School, Glendale, Ohio
Barbara O'Neal, Calhan Elementary School, Calhan, Colorado
Cheryl Pez, Bethany School, Cincinnati, Ohio
Rochelle Rubin, Waterford School District-IMC, Waterford, Michigan
Elizabeth A. Smith, Grayson Elementary School, Waterford, Michigan
Melanie W. Smith, Washington Elementary School, Raleigh, North Carolina
Janet Smith-James, Bartle School, Highland Park, New Jersey
Catherine Snyder, Highland Park School District, Highland Park, New Jersey
Ingrid Snyder, Waterford Village School, Waterford, Michigan
Lee Ann Van Horn, Wake County Public School System, Raleigh, North Carolina
Kathy Wright, Calhan Elementary School, Calhan, Colorado

Reviewers

Marsha Barber, Jefferson County Public Schools, Golden, Colorado
James P. Barufaldi, University of Texas at Austin
Diane Brunner, U.S. Olympic Training Center, Colorado Springs, Colorado
Judy Capra, Jefferson County Public Schools, Golden, Colorado
Candance L. Cline, Etiwanda Elementary School, Etiwanda, California
Larry W. Esposito, University of Colorado at Boulder
Brenda S. Evans, Department of Education, Raleigh, North Carolina
Eva Filsinger, Air Academy High School, Colorado Springs, Colorado
Randy Gray, National Weather Service, Pueblo, Colorado
Leslie Hartten, University of Colorado at Boulder

Steven Holman, Salem, Oregon

Andrew Hudak, University of Colorado at Boulder

Judith Johnson, University of Central Florida, Orlando, Florida

Eric Leonard, The Colorado College, Colorado Springs, Colorado

Ted Lindeman, The Colorado College, Colorado Springs, Colorado

Brownie Linder, Northern Arizona University, Flagstaff, Arizona

Jerry Ludwig, Fox Lane High School, Bedford, New York

Michael J. Madsen, KKTV, Channel 11, Colorado Springs, Colorado

Robert T. Moline, Gustavus Adolphus College, St. Peter, Minnesota

Cherilynn A. Morrow, Space Science Institute, Boulder, Colorado

Rajul Pandya, National Center for Atmospheric Research, Boulder, Colorado

Joseph Pettit, University of Colorado at Boulder

Robert I. Pinney, Colorado Springs, Colorado

Kathleen Roth, Michigan State University, East Lansing, Michigan

Barbara W. Saigo, Saiwood Biology Resources, Montgomery, Alabama

Mary Santlemann, Oregon State University, Corvallis, Oregon

Gail Shroyer, Kansas State University, Manhattan, Kansas

David L. Smith, LaSalle University, Philadelphia, Pennsylvania

Carol Snell, University of Central Florida, Orlando, Florida

John Staver, Kansas State University, Manhattan, Kansas

Joseph Stepans, University of Wyoming, Laramie, Wyoming

Robert Steurer, U.S. West Communications, Colorado Springs, Colorado

Richard Storey, The Colorado College, Colorado Springs, Colorado

Joan Tephly, Marycrest University, Iowa City, Iowa

Lee Vierling, University of Colorado at Boulder

Jack Wheatley, North Carolina State University, Raleigh, North Carolina

Emmet Wright, Kansas State University, Manhattan, Kansas

 Contents

Investigating Weather

Introduction 

Doing Science

Read these questions. Draw a picture to show what you think about each question.

1. What is science?

2. What do people do when they do science?

How Do You Do Science?

Do you wonder? Do you ask questions? Do you try to find answers to your questions? You are doing science!

Look at the following image. It helps us think about doing science.

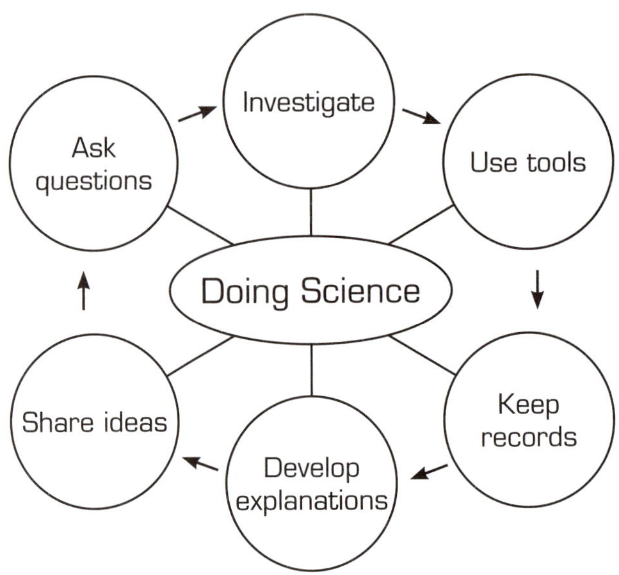

Doing Science Graphic Organizer

People who do science wonder about things. They ask questions. What do you wonder about? What questions do you ask?

Science students ask questions.

Scientists ask questions.

People who do science **investigate**. They find answers to their questions. They **observe**. They use their senses. Their senses help them observe. Do you use your senses to find out about things?

Scientists investigate.

Science students investigate.

People who do science use **tools**. A tool helps do
work. What tools do you think scientists use?
What science tools have you used?

Scientists use tools.

Science students use tools.

People who do science draw pictures. They also write notes. Writing and drawing are ways to keep **records**. Do you keep records?

Scientists keep records.

Science students keep records.

Develop Explanations

People who do science explain their answers.
Their **explanations** are based on what they
have learned. Do you explain things?

Science students develop explanations.

Scientists develop explanations.

People who do science share ideas. Do you share your ideas?

Scientists share their ideas.

Science students share their ideas.

Answering one question may lead to a new question. People who do science are full of questions. Do you have a lot of questions?

Science students ask new questions.

Scientists ask new questions.

Look back at the graphic organizer on page 2. Review the process for doing science. Does the process make sense? Do you think that it is a good way to think about doing science?

Working Together to Do Science

People who do science often work in teams. You will work with a **teammate**. Teammates work together to do a task. Teammates help each other learn.

It is important to practice team skills. You will use five team skills every time you work with your teammate.

■ *Team Skills*

1. Move into your team quickly and quietly.

2. Stay with your team.

3. Speak softly.

4. Share and take turns.

5. Do your job.

1. Move into your team quickly and quietly. When your teacher tells you to meet with your team, find your teammate right away. Do it without talking.

2. **Stay with your team.** Pay attention to your teammate. Work with him or her to do your task. Do not wander around the room.

3. **Speak softly.** Keep your voices down so only
your teammate can hear you.

4. **Share and take turns.** Sharing and taking turns takes practice. Teammates share work. Teammates share ideas.

5. **Do your job.** When you do science, you will do a job. One teammate will be the materials manager. The other teammate will be the messenger. Both jobs are important. You will have a chance to do both jobs. The job descriptions are on pages 15–16.

These descriptions will help you do each job.
There is a special colored wristband to wear for
each job.

The **materials manager** gets the supplies.
Both teammates help clean up. The materials
manager returns the supplies.

The **messenger** may ask another messenger for help. The messenger may ask your teacher for help.

Doing Science with C.Q. and I.O.

Look for C.Q. and I.O. in your student guide.
They will help you do science like a scientist.

Making and Using a Science Notebook or Folder

These pictures will help you remember when to record something. You can record things that happen at other times too.

Doing Science Safely

Doing science is interesting and fun. But you may work with things that can harm you or your clothes. When you see the caution hand, pay attention to the safety hints. Always follow the directions!

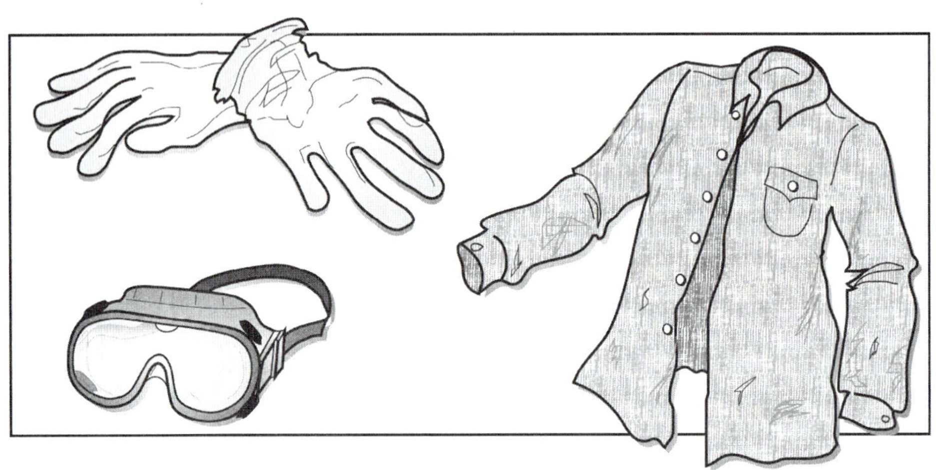

Previewing the Student Guide

Take a tour of your student guide with your teacher.

1. Locate the table of contents. How is a table of contents helpful?

2. Locate the glossary. How is a glossary helpful?

3. Select a lesson.

    a. Look through the lesson looking for titles, headings, and subheadings. What clues do the text sizes give you about the lesson?

    b. Locate the "Team Task," "Team Skills," "Team Jobs," "Team Supplies," and "Directions" sections. How are these things organized to help you learn?

    c. Look for visual clues (photographs, illustrations, charts, graphs, and icons). How can these clues help you learn?

    d. Look for boldfaced words. Why would some words be boldfaced?

4. Look through another lesson. How are the lessons similar? How are they different?

5. How can this guide help you learn? How can it help you to think like a scientist?

You have learned a little about doing science. Now it is time to put on your scientist cap and begin doing science!

Lesson 1

Weather Words

What do you think it is like outside today where C.Q. and I.O. are? What might you do on a day like this?

What do you think it was like outside yesterday where C.Q. and I.O. are?

What might you have done on a day like that?

Finding Out What It Is Like Outside

You listed words that describe what it might be like outside where C.Q. and I.O. are.

What do you think it is like outside where you are? How can you find out?

Team Task

Write words that describe what it is like outside today.

Team Jobs

Materials Manager

Messenger

Team Skills

- Move into your team quickly and quietly.

- Stay with your team.

- Speak softly.

- Share and take turns.

- 1 plastic tray 

- masking tape 

- 1 sheet of writing paper 

- each teammate's notebook or folder 

- 2 pencils 

- 2 wristbands 

1. Before you go outside, do these things to make a writing board.

    a. Turn your tray upside down.

    b. Tape the writing paper to the bottom of the tray.

2. What is it like outside?

   a. Talk about it with your teammate.

   b. Write the words you think of.

   c. How could you tell what it is like outside?
Talk about it.

   d. Be ready to share your team's words and
ideas with the class.

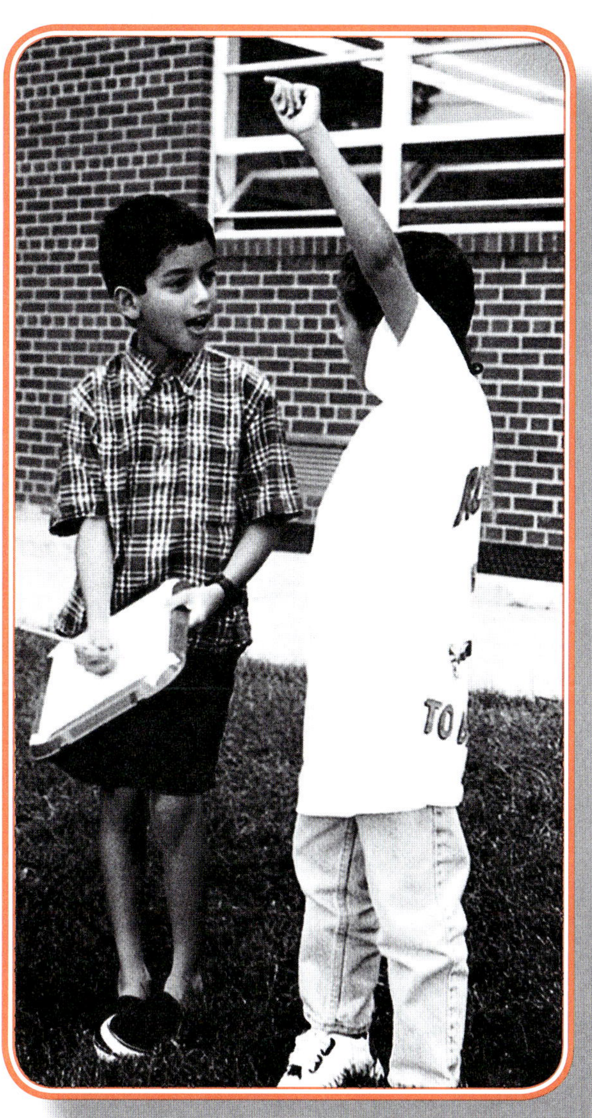

✓ Checking Understanding

Part A: Do these things on your own.

1. Fold a sheet of drawing paper in half.

2. Pick a word from the list of weather words.
Write it on the front of your paper.

3. Inside the folded paper, draw a picture of
what you might do on a day like this.

Part B: Talk about these things with your teammate or class.

1. How were you doing science during this lesson?

2. What activities helped you learn?

3. How did you work as a team?

Lesson 2

Weather Sense

When you think about weather, do you use your senses? Do you use your eyes? Do you use your ears? Do you use your nose? Do you use your skin?

Do you use other things to find out about the weather?

The following poem is about a boy who uses only his senses to find out about the weather. Find out what happens to him!

Dory's Bad Weather Day

The weather turned cold. It snowed all night.

Snow filled the driveway. It was an awesome sight!

"Go out and shovel," Dory's mother demanded,

"We can't get the car out and so we are stranded!"

Dory found himself in a mountain of snow.

The shoveling was hard, boring, and slow.

He looked at the sky and said hopefully,

"The Sun is out now, it feels warmer to me."

"If the Sun stays out and it gets warmer today,

This snow will all melt and I can go play."

The Sun did shine brightly and it felt warmer that day,

But the snow did not melt. It did not go away!

"I could go sledding or have a snowball
fight,

Lie down in the snow and make an angel
just right.

If the snow will not melt, then I have a
plan,

I'll go to the lake and skate if I can!"

But then he remembered how cold it must get,

To skate on the ice without getting wet.

The ice must be solid, a very cold state,

So he kept right on shoveling though he'd much rather skate.

"My eyes saw the Sun shine. The air felt
quite nice.

While the snow would not melt, I could
not trust the ice!

What I saw with my senses was just not
enough.

How can I learn more about weather
stuff?"

Making Your Own Thermometer

Tools can help us learn about weather.
A **thermomete**r is a tool.

You will make a paper thermometer. It will
help you learn to read a real thermometer.

■ *Your Task*

Make a paper thermometer. Learn to read a
thermometer.

- 1 paper thermometer to cut out 

- 1 bottle of glue or glue stick 

- 1 pair of scissors 

- 1 piece of red and white elastic 

- 1 piece of cardboard 

1. Cut out your paper thermometer.

2. Glue the paper thermometer onto your cardboard.

Do not cover the holes in the cardboard.

3. Put the elastic through the holes.

4. Tie the ends of the elastic.

Not too tight!
Not too loose!

Now you have a paper thermometer!

The Hidden Temperatures Game

You have practiced reading your thermometer. Real thermometers measure **temperature**. Temperature is the **degree** of hotness or coldness. Now play a game. You can show what you know about reading a thermometer.

Team Task

Play the Hidden Temperatures game with your teammate. Count how many matches your team makes. Show what you know about reading a thermometer.

Team Jobs

Materials Manager

Messenger

Move into your team quickly and quietly.

Team Supplies

- 2 paper thermometers

- paper to record points

- 2 pencils

- 2 wristbands

1. Decide who will go first.

2. The first player does these things.

    a. Hold the paper thermometer so that your teammate cannot see the numbers on the front.

    b. Move the elastic on your thermometer so that it shows a temperature.

    c. Tell (*but do not show*) your teammate the temperature you picked.

    d. Tell your teammate to find the same temperature on his or her thermometer.

Read your thermometer carefully!

3. Both teammates show the fronts of their thermometers. Do both thermometers show the same temperature?

    a. If your thermometers show the same temperature, give yourselves one point!

    b. If your thermometers do not show the same temperature, find out how they look different. Talk about it until you agree what the thermometers should look like.

How did you do? Show your teacher how many points your team made.

4. Repeat steps 1 through 3 until both teammates have had three or four chances to pick hidden temperatures.

Part A: Read the following sentences. Talk about them with your teammate. Do you think a sentence is true? If so, write "T" on your own paper. Do you think a sentence is false? If so, write "F" on your own paper. Be ready to talk about your choice.

1. There is more than one way to find out about weather.

2. Sometimes you need to know *how* hot or cold it is outside.

3. Thermometers are tools to measure how long something is.

Part B: Talk about these things with your team or class.

1. How were you doing science during this lesson?

2. What activities helped you learn?

3. How did you work as a team?

Lesson 3

Weather That Is Hot! Weather That Is Not!

Finding Out What Is Hot or Not

You shared your ideas about the temperatures of hot water, ice water, and air in the classroom. But what are the **actual** temperatures of those things? "Actual" means "exact."

Investigate to find out.

Team Task

Investigate the temperatures of hot water, ice water, and air. Find out the actual temperatures of hot water, ice water, and air.

Team Jobs

Materials Manager

Messenger

Team Skill

Share and take turns.

- 1 thermometer 

- 2 copies of "What Is Hot or Not Record Page" 

- each teammate's science notebook or folder  

- 2 pencils 

- 2 wristbands 

Take turns measuring the temperature at each center.

1. Do these things as you visit each center with your teammate.

    a. Read the directions.

    b. Follow the directions.

    c. Talk about the questions on your record page.

    d. Write your answers on your own "What Is Hot or Not Record Page."

2. When your teacher tells you to, move to the next center.

3. Be ready to share your findings with the class.

What did you find out about the temperatures of hot water, ice water, and classroom air?

1. What was your question?

Remember, scientists ask questions.

2. How did you investigate?

Remember, scientists investigate.

3. Scientists use tools to collect **data**. "Data" means "information." What tool did you use? What data did you collect?

Remember, scientists use tools.

4. How did you keep records?

Remember, scientists keep records.

5. Did you explain anything? If so, what did you explain?

Remember, scientists explain things.

6. Did you share ideas? If so, how did you share your ideas?

Remember, scientists share ideas.

Measuring Air Temperature

When someone says that it is hot outside today, do you know exactly how hot it is? How could you find out? What tool could you use?

Thermometers can give us data about air temperature. They can show exactly how hot or cold it is outside. Thermometers help us get **accurate** data. "Accurate" means "exact."

Thermometers measure temperature in **degrees**. A degree is a mark on a thermometer, just as an inch is a mark on a ruler.

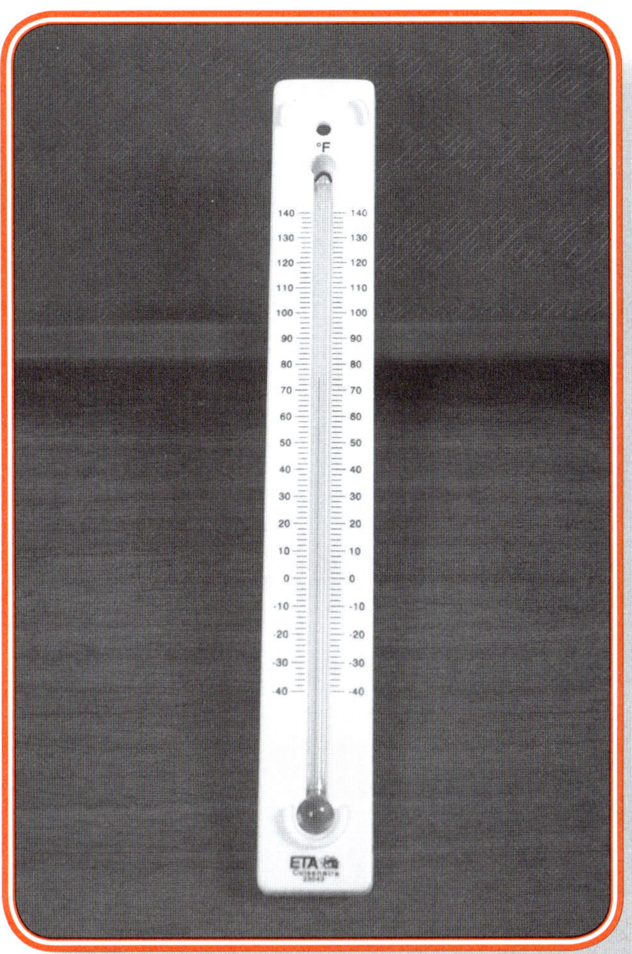

Seventy-five degrees Fahrenheit is the same temperature no matter where it is. Just as 10 inches is the same length no matter where it is. Using a thermometer to show that it is 75 degrees is more accurate than just saying, "It is really hot."

Making Outdoor Observations

How do scientists get data? They observe, measure, and record.

You and your teammate are ready to observe, measure, and record the weather. You will use your senses. You will also use a thermometer.

Team Task

Observe and describe the weather. Measure and record the air temperature.

Team Jobs

Materials Manager

Messenger

Team Skill

Share and take turns.

- 1 thermometer 

- 2 copies of "Weather Watch Record Page—
  Lesson 3" 

- each teammate's science notebook or folder 

- 2 pencils 

- 2 wristbands 

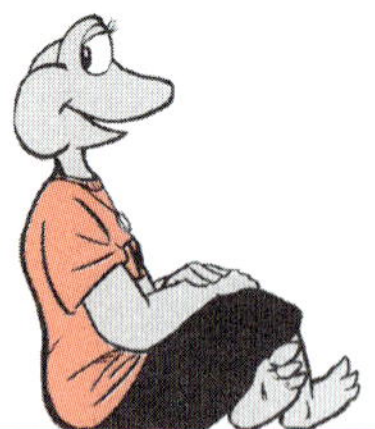

Do these things every time you observe the weather and measure the temperature outside.

1. Complete your record page.

a. Write today's date.

b. Circle "morning" or "afternoon."

2. Use your senses.

a. What is the weather like today?

b. Circle your answer.

3. Use your thermometer.

   a. Hold the thermometer in the air. Do not touch the glass tube.

   b. Read the thermometer. Both teammates should read it.

   c. Agree on the temperature.

   d. Record the temperature.

Talk about these questions with your teammate. Be ready to share your ideas with the class.

1. What can you find out about the temperature by using your senses?

2. What can you find out about the temperature by using a thermometer?

3. Why is it sometimes important to describe the temperature accurately?

Part A: On your own, do these things. Write your answer in your science notebook or on a piece of paper to put in your folder.

1. Pick a temperature. Write the number at the top of a piece of paper. If it were this temperature outside,

   a. what might you do?

   b. what might you wear?

   c. what might you eat?

   d. what might you drink?

2. Draw your answers to the questions in step 1. Write at least two sentences explaining them.

Part B: Talk about these things with your teammate or class.

1. How were you doing science during this lesson?

2. What activities helped you learn?

3. How did you work as a team?

Lesson 4

Windy Weather

Have you ever stood out in the **wind**? What is wind? Wind is moving air.

What Can Your Senses Tell You about Wind?

What data can you collect about the wind using only your senses?

■ *Your Task*

Use your senses to collect data about the wind. Describe the wind using your data.

■ *Your Supplies*

- 1 copy of "Finding Out about Wind Record Page"

- your science notebook or folder

- 1 pencil

1. Observe the wind by using your senses. Draw or write what you find out. 

   a. What do you see?

   b. What do you hear?

   c. What do you feel?

   d. What do you smell?

2. Compare your data with your teammate's data.

    a. How are your data the same?

    b. How are your data different?

What Can Science Tools Tell You about Wind?

Wind Speed

Scientists are interested in wind. They care about **wind speed**. Wind speed is the strength of the wind.

Scientists can describe wind using a **wind scale**. A wind scale helps them tell how fast the wind is blowing. Other tools help them measure exact wind speeds.

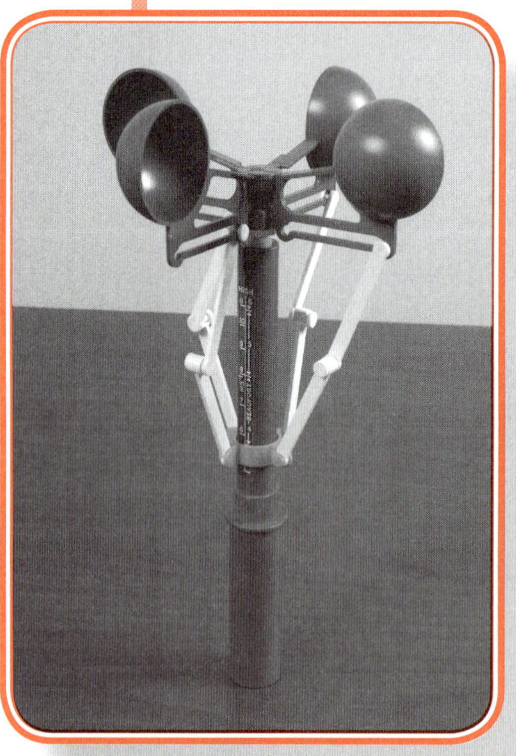

You can use a scale to help you describe wind. If you have a flagpole, you can use the following wind scale.

Flag Speed Chart

	Flag is barely moving or completely still.	Calm
	Flag is gently swaying or flapping.	Breezy
	Flag is flapping furiously.	Windy

What do you think this tool can tell you about the wind? It is an **anemometer**. An anemometer can measure wind speed in miles per hour. An exact wind speed is more accurate than saying that it is windy.

Scientists are also interested in **wind direction**. Wind direction is the direction from which it blows.

Scientists use a **wind vane** to find wind direction. A wind vane tells the direction of the wind.

A wind vane and wind speed indicator are part of this **weather station**. A weather station can measure several things at once. Can you find the wind vane? Can you find the wind speed indicator?

These tools give us data about the wind. The data can help us describe weather more accurately.

Making Outdoor Observations

In lesson 3, you used a thermometer to measure temperature. Now you will use a weather station. You will investigate weather just like scientists!

Team Task

Observe and describe the weather. Measure and record the temperature, wind direction, and wind speed.

Team Jobs

Materials Manager

Messenger

Team Skill

Share and take turns.

- 1 weather station (shared)

- 1 thermometer

- 2 copies of "Weather Watch Record Page—Lesson 4"

- each teammate's science notebook or folder

- 2 pencils

- 2 wristbands

1. Complete your record page.

    a. Write today's date.

    b. Circle "morning" or "afternoon."

Do these things every time you measure temperature and wind.

2. Use your senses.

    a. What is the weather like today?

    b. Circle your answer.

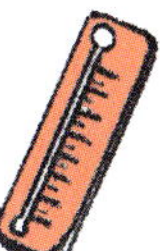

3. Use your thermometer.

    a. Hold the thermometer in the air. Do not touch the glass tube.

    b. Read the thermometer. Both teammates should read it.

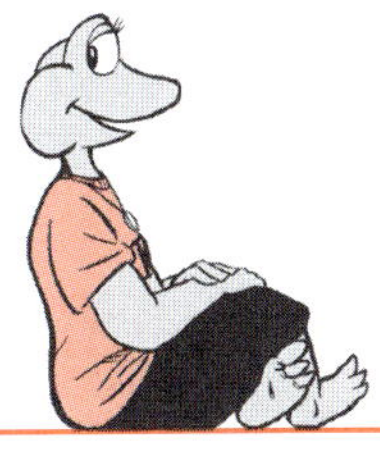

Wait until the red liquid has stopped moving.

    c. Agree on the temperature.

    d. Record the temperature.

4. Observe your team's weather station.

    a. Read the wind vane.

    b. Record the direction that the wind vane is pointing.

    c. Read the wind speed indicator.

    d. Record the wind speed.

Talk about these questions with your teammate. Be ready to share your ideas with the class.

1. What can you find out about the wind by using your senses?

2. What can you find out about the wind by using a wind vane and wind speed indicator?

3. Why is it important to describe the wind accurately?

Part A: On your own, do these things. Write your answer in your science notebook or on a piece of paper to put in your folder.

1. Pick a wind speed. Write the number at the top of a piece of paper. If the wind were blowing this fast,

    a. what might you do?

    b. what might you wear?

2. Draw your answers. Write at least two sentences to describe your drawing.

Part B: Talk about these things with your teammate or class.

1. How were you doing science during this lesson?

2. What activities helped you learn?

3. How did you work as a team?

Lesson 5

Wet Weather

How much rain makes a gully washer? How much rain has to fall for it to be raining cats and dogs?

What other ways do people describe rain? How can you tell if it is raining a lot or a little?

Are There Different Kinds of Precipitation?

Rain is a kind of **precipitation**. Precipitation is any form of water that falls to Earth from clouds.

What kind of precipitation is this?

What kind of precipitation is this?

Can You Measure Precipitation?

What do you think? What might a tool that measures precipitation look like?

Weather scientists use different tools to
measure different kinds of precipitation.
They use a **rain gauge** to measure rain.
A rain gauge is a tool that collects rain in
a tall, thin cup.

What do you think this tool can show you
about rain?

A rain gauge collects rain in a container that can be marked in inches. Snow is measured in inches too! Does it snow where you live? If it does, you can use a ruler or a yardstick to find out how much snow has fallen.

Making Outdoor Observations

So far, you have used a thermometer to measure temperature. You have used a wind speed indicator to measure wind speed. You have used a wind vane to measure wind direction. Now you will use a rain gauge or ruler to measure precipitation. Just like a weather scientist, you will record the data you collect.

Team Task

Observe and describe the weather. Measure and record the temperature, wind direction, wind speed, and amount of precipitation.

Team Jobs

Materials Manager

Messenger

Team Skill

Share and take turns.

- 1 weather station (shared)

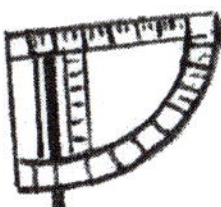

- 1 thermometer

- 2 copies of "Weather Watch Record Page—Lesson 5"

- each teammate's science notebook or folder

- 2 pencils

- 2 wristbands

1. Complete your record page.

    a. Write today's date.

    b. Circle "morning" or "afternoon."

Do these things every time you measure temperature, wind, and precipitation.

2. Use your senses.

    a. What is the weather like today?

    b. Circle your answer.

3. Use your thermometer.

   a. Hold the thermometer in the air. Do not touch the glass tube.

   b. Read the thermometer. Both teammates should read it.

Wait until the red liquid has stopped moving.

   c. Agree on the temperature.

   d. Record the temperature.

4. Observe your team's weather station.

    a. Read the wind vane.

    b. Record the direction that the wind vane is pointing.

    c. Read the wind speed indicator.

    d. Record the wind speed.

    e. Read the rain gauge or ruler.

    f. Record the amount of rain or snow.

Sharing Ideas

Talk about these questions with your teammate. Be ready to share your ideas with the class.

1. What can you find out about precipitation by using your senses?

2. What can you find out about precipitation by using tools?

3. Why is it important to describe the amount of precipitation accurately?

Investigating Water in a Closed Dish

You have measured precipitation. Where does precipitation come from? You will investigate to find out!

■ Team Task

Observe and describe what happens to water in a closed dish.

■ Team Jobs

Materials Manager

Messenger

■ Team Skill

Stay with your team.

- 1 tray

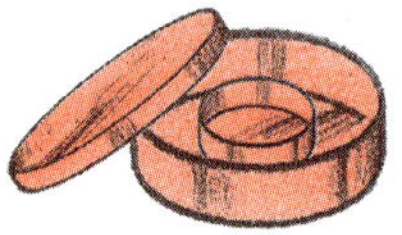

- 1 closed dish (with 2 rings inside)

- 1 medicine cup full of water

- 1 piece of masking tape

- 2 copies of "Investigating Water in a Closed Dish"

- each teammate's science notebook or folder

- 2 pencils

- 2 wristbands

1. Write today's date and your name on your record page. 

2. Make sure that your team tray has one dish (top and bottom) and one medicine cup full of water.

3. Write your names on a piece of masking tape. Stick it to your tray.

4. Your team is ready to go outside.

1. Open your dish.

2. Carefully pour the water from the medicine cup into the outer circle of the dish. It is very important not to pour any into the inner circle.

3. Close the dish.

4. Write and draw to explain what you observe.

5. Leave your investigations in an area where the Sun will warm them all day.

1. Return to your investigation outside. Check the names on the tray to be sure it is your team's investigation.

2. Write and draw to explain what you observe. 

   a. Is the water in the same place as it was earlier in the day? Where is the water?

   b. Write to record your observation.

3. When your teacher instructs you to go inside, clean up your materials.

The Water Cycle

Water can be found in three ways on our planet. Can you think of the three ways? Water can be a liquid, a solid, or a gas.

Liquid water is the way most people think about water. Something that you can pour into a glass would be a liquid.

Solid water is something familiar to most of you also. Solid water is ice. You probably put ice in your drink in the summer time to cool it off.

The last way we find water is not as easy to think about. Water can be found as a gas. What is a gas? A gas is something that we cannot see. Water gas is completely invisible, but it is there! What does all of this have to do with weather? Read on your own or with a partner to find out. Use your "Water Cycle" handout to help you remember things about the words in the boxes.

We can find water almost everywhere!

Water is on the ground.

It is in the rivers.

We can see it in the rivers; it is liquid
water.

It is in the lakes.

We can see it in the lakes; it is liquid water.

It is in the oceans.

We can see it in the oceans; it is liquid water.

Water is in the sky.

It is in the air.

We *cannot* see it in the air; it is water gas.

How does the water get from the ground to the sky?

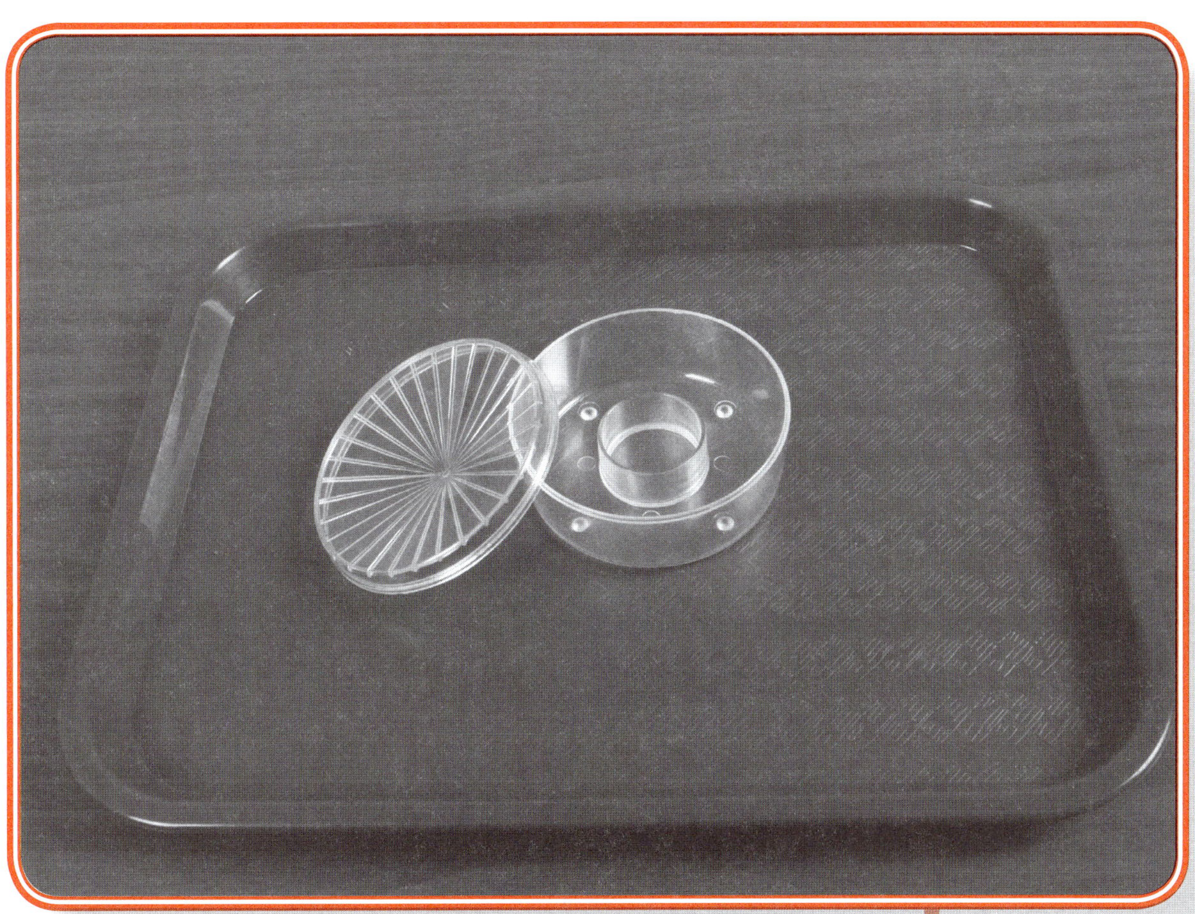

The Sun warms the water on the ground.

When the water gets warm, it changes the liquid water into water gas.

This change is called **evaporation**.

Remember, we cannot see the water gas. Gas is invisible.

As the water in the air cools, it changes the water gas into tiny drops of liquid water.

This change is called **condensation**.

The tiny water droplets form a cloud.

As the tiny water droplets group together,
they make bigger drops.

The bigger drops fall out of the sky.

This is called precipitation.

When the precipitation reaches the ground, it goes into puddles, lakes, rivers, or oceans.

This is called **accumulation**.

Evaporation, condensation, precipitation, accumulation! It is called the **water cycle**, and it happens all the time.

Investigating Water in a Plastic Bag

You have investigated water in a closed dish. You have read about the water cycle. Let's see what happens to water in a plastic bag.

■ Team Task

Observe and describe what happens to water in a closed plastic bag.

■ Team Jobs

Materials Manager

Messenger

■ Team Skill

Stay with your team.

- 2 resealable plastic bags

- 1 metric ruler

- water

- 2 copies of "Investigating Water in a Plastic Bag"

- strong tape

- each teammate's science notebook or folder

- 1 permanent marker

- 2 pencils

- 2 wristbands

1. Write today's date and your name on your record page.

2. Write your name on your plastic bag.

3. Draw a line 3 centimeters from the bottom all the way across your bag.

4. Draw a cloud at the top of your bag.

5. Put water in your bag. Stop when you get to
the line that you drew.

6. Zip your bag closed.

7. Now your team is ready to go outside to set
up.

1. Get a piece of strong tape from your teacher.

2. Carefully help your teammate tape the top of the bag to the wall.

3. Write and draw to explain what you observe.

4. Write and draw what you **predict** will happen.

5. Leave your investigations taped to the sunny wall where the Sun will warm them all day.

1. Return to your investigation outside. Check the names on the bag to be sure it is your investigation.

2. Write and draw to explain what you observe. 

   a. Is the water in the same place that it was earlier in the day?

   b. Is the water on the sides of the bag?

3. Touch the side of the bag. What happens to the water on the side of the bag? Write and draw to record your observation. 

4. When your teacher instructs you to go inside, clean up your materials.

Part A: On your own, do these things. Write your answer in your science notebook or on a piece of paper to put in your folder.

1. Pick a type and amount of precipitation. Write the type and amount at the top of a piece of paper. If this much precipitation had fallen,

   a. what might you do?

   b. what might you wear?

2. Draw your answers. Write at least two sentences to describe your drawing.

3. Draw a picture to show what you know about the water cycle. Label your picture.

Part B: Talk about these things with your teammate or class.

1. How were you doing science during this lesson?

2. What activities helped you learn?

3. How did you work as a team?

Lesson 6

Weather Reports

When scientists find out about something, they share! They share their data to help explain what they found out. These data are called **evidence**. Evidence is what you find out that helps you know something.

Making a Weather Report

Get ready to share what you have found out about weather.

■ *Team Task*

Use your data to make a weather report. Present your team's weather report to the class.

■ *Team Jobs*

Materials Manager

Messenger

■ *Team Skill*

Ask for help and give help.

- 1 copy of "Weather Report Record Page"

- each teammate's notebook or folder

- 2 pencils

- 2 wristbands

1. Pick two days from the "Class weather chart."

a. Make sure the two days are together.

b. Copy the data for the two days onto your "Weather Report Record Page."

Class Weather Chart

70°	55°	50°	45°	55°
Monday	Tuesday	Wednesday	Thursday	Friday

<table>
<tr><td colspan="5" align="center">Class Weather Chart</td></tr>
<tr><td>70°
Monday</td><td>55°
Tuesday</td><td>50°
Wednesday</td><td>45°
Thursday</td><td>55°
Friday</td></tr>
<tr><td></td><td></td><td></td><td></td><td></td></tr>
<tr><td></td><td></td><td></td><td></td><td></td></tr>
<tr><td></td><td></td><td></td><td></td><td></td></tr>
</table>

2. Use the data to describe the weather. You might tell about these things.

a. How much hotter or colder the air temperature was on the first day than the second day

b. How the wind changed direction from the first day to the second day

c. How the wind speed changed from the first day to the second day

d. How the kind of precipitation changed from the first day to the second day

e. How the amount of precipitation changed from the first day to the second day

f. How the sky changed from the first day to the second day

3. Do something fun to present your weather report!

a. You could pretend to be television weather forecasters.

b. You could make a poster.

c. You could put on a play.

Weather Changes

You collected weather data and found out that weather can change from day to day. You shared your evidence as scientists do!

You talked about ways that weather can change from season to season. Share your evidence. Show how the weather changes from season to season.

The Secret Season Game

Playing a game is one way to show what you know about something. Use the Secret Season game to show what you know about weather changes from one season to the next.

■ *Team Task*

Play the Secret Season game with your teammate.

■ *Team Jobs*

There are no jobs for this activity.

■ *Team Skill*

Ask for help and give help

■ *Team Supplies*

There are no team supplies for this activity.

Directions

1. One teammate picks a secret season from the secret season box.

2. Give your teammate clues to help him or her guess the secret season. To give clues, read these sentences. Finish each sentence with a word that gives a clue about the secret season.

    a. During this season, the temperature would probably be _______________.

    b. During this season, the wind would be _______________.

    c. During this season, the kind of precipitation that would fall would be _______________.

    d. During this season, the amount of precipitation would be _______________.

e. During this season, I would wear a

   ______________.

f. During this season, I would do

   ______________.

g. During this season, I would eat

   ______________.

3. When your teammate guesses the secret season, it is his or her turn to pick a secret season from the box.

4. Repeat step 2.

5. Play the Secret Season game until both teammates have had at least two turns.

Part A: Share more ideas about how the weather changes over the year. Do these things.

1. Exchange your secret season card with a classmate.

2. When it is your turn, put your secret season card on the class calendar.

3. Be ready to tell why you put your secret season card where you did.

Part B: Talk about these things with your teammate or class.

1. How were you doing science during this lesson?

2. What activities helped you learn?

3. How did you work as a team?

Lesson 7

Sudden Changes

Weather changes every day. Sometimes the changes are slow. Sometimes the changes are fast. Sometimes the changes are small. Sometimes the changes are big.

I.Q.

What do you think just happened in these pictures?

Storm Pictures

One kind of storm that happens everywhere is a thunderstorm. A thunderstorm is a storm that involves thunder and lightning. Almost everyone has had a thunderstorm happen nearby. Think of a time when you observed a thunderstorm.

Your Task

Make a storm picture that shows a thunderstorm.

Your Supplies

- drawing paper

- markers or crayons

1. Think of a time that a thunderstorm happened near you.

2. Draw a picture that shows what you observed during the storm.

   a. Show what you saw.

   b. Show what you heard.

   c. Show what you felt.

3. Show yourself in the picture.

4. Be ready to share your storm picture with the class.

The Science of Thunderstorms

Compare your ideas about thunderstorms with what scientists know.

Big clouds show up in the sky before a thunderstorm. The clouds start to look dark. The wind begins to blow. Sometimes the wind blows really hard.

Lightning is part of all thunderstorms. Lightning is a flashing of light caused by electricity passing from one cloud to another cloud. Lightning can also occur between a cloud and the ground.

Lightning is very dangerous. It can start fires. It can hurt or kill people and animals! If you see lightning, get inside.

Thunder is the loud crashing or rolling noise that follows lightning. Thunder is caused by lightning. If you hear thunder, get inside.

Thunderstorms usually include rain or
hail. Hail is chunks of ice made in clouds.
Some hail is small. It can be the size of a
pea. Some hail is big. Sometimes people
talk about hail the size of golf balls.

If it rains a lot, a **flood** can happen. A **flash flood** is one kind of flood. A flash flood can happen in just a few minutes. Water fills streets, ditches, and streams.

A flash flood can be very dangerous. Be very careful if you hear flash flood warnings!

Storm Stories

Your storm picture tells the story of a time you observed a thunderstorm. Read the three storm stories that follow. Then tell what kind of storm you think each story is about.

Carmen's Storm Story

Carmen sat in her classroom. She was looking out the window. She was observing the sky. The clouds looked black. She was listening to the sky. It was rumbling.

Carmen's teacher told everyone to line up at the door. Everyone walked out into the hall. They faced the wall. They got down on their knees. They put their arms over their heads and curled up like a ball.

Carmen heard a loud roar. Glass was breaking. Her arms tightened around her head.

The loud roar did not last very long. Soon the sky was quiet again. Mrs. Foster let the children get up. "The storm is over," she said. "You were great listeners. You moved quickly and quietly. We are all safe because we did the right thing."

1. What kind of storm do you think happened near Carmen's school? Why do you think so?

2. How do you think the temperature might have changed?

3. How do you think the wind might have changed?

4. How to you think the precipitation might have changed?

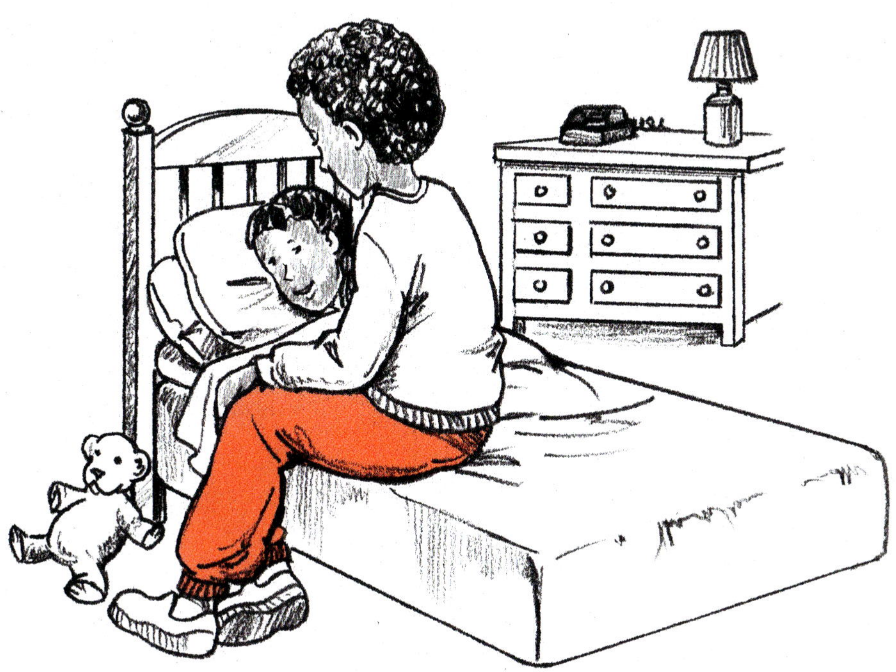

Lakeisha's mom woke her up early one morning. Lakeisha was tired. She did not want to get up. Her mom said that they needed to go to Grandma's house. A storm was on its way.

Lakeisha lives near the beach. Her grandma lives farther away from the ocean. Lakeisha's mom said that the storm would hit hardest close to the ocean. They would be safe at Grandma's house.

The car was packed. They were ready to go. Lakeisha noticed that the windows on the house were covered as they drove away.

When they got to Grandma's house,
Lakeisha and her mom watched the news.
It showed pictures of the storm. The wind
was bending trees. The rain was making
rivers overflow. The streets looked like
rivers. Trucks and cars were washing
away as if they were toys. A big boat was
laying upside down on somebody's lawn.
Ocean waves tore down houses as the
water crashed into the land.

Lakeisha hugged her mom. She was glad
that they were safe.

1. What kind of storm do you think happened near Lakeisha's house? Why do you think so?

2. How do you think the temperature might have changed?

3. How do you think the wind might have changed?

4. How to you think the precipitation might have changed?

Jay was excited about the snow. He
wanted to go outside and play. His mother
said that he should stay inside. Jay sat
by the window watching the flakes float to
the ground.

Soon the windows started to shake. The wind made whistling sounds. It started snowing very hard. The whole outside disappeared in a blanket of white.

1. What kind of storm do you think happened near Jay's house? Why do you think so?

2. How do you think the temperature might have changed?

3. How do you think the wind might have changed?

4. How do you think the precipitation might have changed?

The Science of Storms

Compare your ideas about the storms described in the stories with what scientists know about storms.

Tornadoes

Tornadoes are very dangerous storms. They are columns of spinning air. They form in storm clouds. They can sound like a roaring train.

Tornadoes can happen during spring or summer storms. If they touch the ground, they can cause a great deal of damage. They are sometimes strong enough to knock down buildings and move cars.

Hurricanes are big storms that form over oceans. The storm clouds move in a circle. They have strong winds. The "eye" of the storm is in the center of the spinning clouds. There is no wind in the eye.

Hurricanes usually occur between June and November. They can bring huge amounts of rain, flooding, and high winds. They can wash away and destroy anything in their paths.

Blizzards are storms caused by blowing snow. It is hard to see during a blizzard. Blizzards can bring several feet of snow. It can get 3 or 4 feet deep! Air temperatures can drop below freezing. Freezing temperatures are dangerous.

Blizzards usually happen in winter or spring. Deep snow can make it hard to drive. Heavy snow can damage power lines, trees, and buildings.

Meteorologists are weather scientists. They study weather. They use their senses. They use tools like the ones you used. They also use special tools.

Doppler radar and satellites are special tools that scientists use. The tools help them watch for storms. They warn people about dangerous storms. Do you watch weather reports?

Discuss the following questions with your teammate and class.

1. What kinds of storms do you experience where you live?

2. What kinds of things should you do to stay safe during these storms?

✓Checking Understanding

Part A: Do this exercise with your teammate. Be ready to share your storm information with the class.

1. Talk about one of the storms you read about.

2. Be ready to tell the class how the weather would have to change from a nice day to be like your storm.

Part B: Talk about these things with your teammate or class.

1. How were you doing science during this lesson?

2. What activities helped you learn?

3. How did you work as a team?

Weather Wise

Sometimes using your senses can tell you enough about the weather. Sometimes you need to use a tool.

Senses or Tools?

You will read a story about Keiko. You will decide when you can use senses. You will decide when you need a tool.

■ Your Task

Read the story. At certain spots, decide whether you would use your senses or a tool.

■ Your Supplies

- 1 copy of the "Senses or Tools?"

- your science notebook or folder

- 1 pencil

1. Read or listen to the story.

2. When you reach a stopping place in the story, stop and think. 

3. Circle your answer on your record page. 

4. Be ready to explain your answer.

Keiko woke up early Saturday morning. She kicked off her covers and jumped out of bed.

Keiko was in a hurry to find out if she could go to the beach. Her mother told her that she could go if it was sunny and warm outside.

Keiko hoped it was not cloudy or rainy outside. How could she find out?

1. Stop and Think 

   a. Would she use her senses or a tool?

   b. If she would use a tool, which tool?

   c. Mark your record page.

Hooray! It was sunny outside!

It was sunny, but was it warm enough to go to the beach? Keiko needed to tell her mom just how warm it was outside.

How could Keiko find out how warm it
was outside?

2. Stop and Think 

a. Would she use her senses or a
tool?

b. If she would use a tool, which
tool?

c. Mark your record page.

Hooray! Keiko danced around her room. It was warm enough to go to the beach!

Keiko took a big bag out of her closet. She packed all the things she wanted to take to the beach.

Keiko wanted to take her new sport kite. She knew that there had to be some wind blowing for her to fly her kite. But the package said not to fly the kite if the wind was blowing too fast. The package said there should be a moderate breeze.

How could Keiko find out if the wind was
blowing too fast?

3. Stop and Think 

    a. Would she use her senses or a
tool?

    b. If she would use a tool, which
tool?

    c. Mark your record page.

Hooray! The wind was blowing just
enough for Keiko to fly her new kite.

Keiko and her mother had a wonderful
day at the beach. Then the weather
changed. Keiko thought it might rain.

What made Keiko think it was going to rain?

4. Stop and Think 

   a. Would she use her senses or a tool?

   b. If she would use a tool, which tool?

   c. Mark your record page.

Keiko was right! It started to rain. Keiko and her mom hurried home.

The sky turned darker and darker.

"This is going to be some storm!" Keiko's mother exclaimed as they hurried inside their house.

"I wonder how much rain will fall," Keiko said to her mother.

How could Keiko find out how much rain
fell?

> **5.** Stop and Think 
>
> **a.** Would she use her senses or a
> tool?
>
> **b.** If she would use a tool, which
> tool?
>
> **c.** Mark your record page.

Part A: Discuss these questions with your teammate. Write or draw your own answers.

1. Why do scientists use tools when they describe the weather?

2. When might a scientist use weather tools?

3. You need to know about the weather too! When are some times that you might need to use a tool instead of just your senses? Why?

Part B: Talk about these things with your teammate or class.

1. How were you doing science during this lesson?

2. What activities helped you learn?

3. How did you work as a team?

Glossary

accumulation: When precipitation reaches the ground, it goes into puddles, lakes, rivers, or oceans.

accurate: Exact.

actual: Exact.

anemometer: A tool to measure wind speed.

blizzard: A storm caused by blowing snow.

condensation: As water in the air cools, it changes the water gas into tiny drops of liquid water.

data: Information collected in a scientific investigation.

degree: A mark on a thermometer, just as an inch is a mark on a ruler.

evaporation: When water gets warm, it changes the liquid water into water gas.

evidence: What you find out that helps you know something.

explanation: A statement based on what you have learned.

flash flood: A flood that happens quickly.

flood: Water covering the land.

hail: Chunks of ice made in clouds.

hurricane: A storm with terrible wind and rain that forms over oceans.

investigate: The way to find answers to questions.

lightning: A flash of light caused by electricity passing from one cloud to another cloud.

materials manager: The person who has the team job of getting the supplies that are listed in the "Team Supplies" section for each lesson; when the team task is completed, the materials manager returns the supplies to the supply table.

messenger: The person who has the team job to ask another team's messenger or your teacher for help if the team gets stuck.

meteorologist: A scientist who studies weather.

observe: To use senses to gather information.

precipitation: Any form of water that falls to Earth from clouds.

predict: To write a statement of what you think might happen.

rain gauge: A tool to measure rain.

record: A place where a scientist keeps track of his or her data, including writings and drawings.

teammate: A person who is part of a team working together to complete a task.

temperature: The degree of hotness or coldness of something.

thermometer: A tool to measure temperature.

thunder: The loud crashing or rolling noise that follows lightning.

thunderstorm: A storm that involves thunder and lightning.

tool: A tool helps you do work.

tornado: A dangerous storm that involves a column of spinning air.

water cycle: The movement of water through evaporation, condensation, precipitation, and accumulation.

weather station: A place where weather data is recorded.

wind: The movement of air.

wind direction: The direction from which wind blows.

wind scale: A tool that helps describe how fast the wind is blowing.

wind speed: The strength of the wind.

wind vane: A tool to measure wind direction.

Acknowledgments

Photo Credits

Carlye Calvin: p. 28; p. 29 (top); p. 29 (bottom); p. 45; p. 46; p. 130; p. 134

Comstock: p. 101

Copyright University Corporation for Atmospheric Research: p. 4 (right); p. 5 (right); p. 131; p. 149

Craig Daniel Wood/Visuals Unlimited: p. 81

Greg Stumpf: p. 146

iStockphoto: p. 3 (right); p. 8 (right); p. 9 (right); p. 31; p. 65; p. 67; p. 69 (top); p. 69 (middle); p. 69 (bottom); p. 71; p. 97; p. 98; p. 102; p. 105; p. 106; p. 148

NASA/TSADO/Tom Stack & Associates: p. 147

PhotoDisc: p. 7 (right); p. 80; p. 82; p. 99; p. 100; p. 104

Shutterstock: p. 1; p. 3 (left); p. 4 (left); p. 6 (left); p. 6 (right); p. 133

Special thanks to the administration, teachers, students, and parents of John Adams Elementary School, Colorado Springs, Colorado, for allowing us to photograph students "doing science": p. 7 (left); p. 11; p. 12; p. 15; p. 16